Engineering for Inclusion 2019

Held at the 2019 AIChE Annual Meeting

Orlando, Florida, USA
10-15 November 2019

ISBN: 978-1-7138-0560-1

Printed from e-media with permission by:

Curran Associates, Inc.
57 Morehouse Lane
Red Hook, NY 12571

Some format issues inherent in the e-media version may also appear in this print version.

Printed with permission by Curran Associates, Inc. (2020)

For permission requests, please contact AIChE
at the address below.

AIChE
120 Wall Street, FL 23
New York, NY 10005-4020

Phone: (800) 242-4363
Fax: (203) 775-5177

www.aiche.org

Additional copies of this publication are available from:

Curran Associates, Inc.
57 Morehouse Lane
Red Hook, NY 12571 USA
Phone: 845-758-0400
Fax: 845-758-2633
Email: curran@proceedings.com
Web: www.proceedings.com

TABLE OF CONTENTS

582663 RAPID Manufacturing Institute and AIChE

Process intensification is of worldwide interest as a type of advanced manufacturing. It aims for new, more compact, and more efficient processes, especially by applying component integration, novel driving forces, and modular deployment of these units. Join the PAIC Town Hall to hear about the leadership role AIChE plays via RAPID: a US Department of Energy advanced-manufacturing institute for process intensification and the development of process and operation principles and their applications.

582664 The Importance of Credentialing and Licensure

How do additional credentials strengthen an engineer's competitive position? What are the credentials that a chemical engineer may find helpful? How does a credential differ from additional education, licensure and advanced degrees? Information will be shared that will hopefully open up additional avenues for engineers of all ages to consider in their career planning.

582665 Legal Developments on Plastics: Basel Convention and the Circular Plastics Economy

The PAIC Legal Developments Subcommittee follows legal issues arising in: energy, food, water and other resources; climate change; chemical regulation; anti-terrorism, physical and cybersecurity, and chemical process safety; ethics, manufacturing and business law. At the PAIC Town Hall, we will review the May 2019 addition of plastic waste to the Basel Convention and its provisions regarding the development of a worldwide circular plastics economy; and the progress of the Expert Working Group pursuant to the Convention.

582666 Responding to Paradigm Shifts

Paradigm shifts generally lead to winners and losers. Manufacturing, chemicals, and energy illustrate three important areas of change driven by technology innovation and a shift from product to integrated system thinking. At the PAIC Town Hall, we will discuss the need for new frameworks that mesh technology, societal, and environmental dimensions and for new ways to respond to on-going shifts.

588861 The Chemical Engineering Profession and Its Infinite Possibilities

Tayo Femi-Fowode is a passionate Chemical Engineer with over 10 years of corporate and leadership experience. He is currently a Business Integration Leader for GE Power's Monitoring and Diagnostic Center in Atlanta, GA. In his role, Tayo drives digital integration of 5000 monitored turbines, generators and other equipment to minimize downtime, optimize plant performance and save operating cost for more than 950 power plants in 75 countries and serving 350 million people worldwide. Before joining GE Power in 2011, he was Project Manager in the asset management division of Rolls-Royce Energy.

A second-generation Nigerian-American, Tayo holds an MBA from Columbia Business School with concentration in Strategy and Finance. He also holds a Master of Science and a Bachelor of Science *(magna cum laude)* degrees in Chemical Engineering from Columbia University and Prairie View A&M University respectively.

588862 Direct contact membrane distillation (DCMD) membrane distillation techniques for desalination

Direct contact membrane distillation (DCMD) is one of the simplest and most widely used membrane distillation techniques for desalination. It is a thermally driven separation process and is sought to achieve high water vapor flux while being thermally efficient. Typically, a hydrophobic porous membrane is used and water vapor from the salt solution is allowed to pass through the pores and condense on the other side of the membrane. Investigation on other structures of membranes is carried out. Membranes include those in the form of an integral composite, in which the structures of PVDF membranes are changed, and a stacked composite, in which different membranes are stacked on top of each other (e.g. a hydrophobic ePTFE membrane over a hydrophilic PVDF membrane, and a hydrophobic PVDF membrane over another hydrophobic PVDF membrane). These membranes were found to yield a higher water vapor flux than the conventional single hydrophobic membrane. Membranes are also characterized using a variety of characterization techniques: Contact angle, liquid entry pressure (LEP), bubble-point pressure, scanning electron microscopy (SEM), and atomic force microscopy (AFM).

588863 My Career Path

My career thus far has been shaped by many factors learned early in life. It started in my hometown of Yazoo City, Mississippi. It was there that I had the job that I consider to be my first engineering experience which was working with my dad at Gainwell Tire Services. While there, I discovered that I had a fondness for hands-on work and gravitated toward working closely with people as part of a team. This was instrumental in not only my early career development but shaped my life as a whole. From there, while in high school, I worked at our local construction company, Parker Builders. During this time, I developed a strong partiality for repairing process equipment. Also, a sense of work ethic was formed as there were several days where I worked on a hot tin roof for hours. After graduation, I headed to college at Louisiana Tech University. While there, I was a part of the football team while pursuing my Chemical Engineering degree. This conceivably impossible task was made possible largely due to the work ethic learned early in life. It also helped further developed my team-membership skills.

Later, I transferred to Southern Arkansas University. It was here that I not only compounded upon my team-membership skills but developed into a team leader as a captain of the football team. While there I took on several internships which included stints in the chemical, aerospace, energy, and mechanical fields. One of those internships was Lanxess Solutions in El Dorado, Arkansas. The skills learned early in life helped me very much in many of the projects given during this time and were good enough to be hired full time as a Shift Leader upon graduating college. In this role, I developed many technical skills and gained an understanding of the operations side of the plant. After some time there, I went to New Orleans, Louisiana to work at Galata Chemicals as a Process Engineer. While at Galata, I was able to further develop technical skills as well as learn the safety aspect that is imperative in a plant setting. After spending a year and a half there, I was offered a Process Engineer II position at Oxea Chemicals, my current role. Thus far I have learned even more about the chemical field and continue to grow professionally with the organizational training offered.

588932 Sustainable Energy Drivers: A Perspective from a Career in the Nuclear Sector

Dr. Harold T. Conner, Jr. has a 52-year career in the nuclear sector, having determined at an early age that he wanted a career in chemical engineering. Along the way, he blazed an impressive path by achieving a lot of firsts: the first African American student in the University of Tennessee-Martin Engineering School, the first cooperative engineering student at the University, and the first African American to receive bachelor's and master's degrees in chemical engineering from the University of Tennessee-Knoxville. He worked for more than 30 years in Oak Ridge, working in almost every facility at the Oak Ridge K-25 Site, a uranium enrichment facility that would be closed down in the mid-1980s and then undergo cleanup. Conner became site manager of the K-25 Site and was then named vice president of Environmental Management and Enrichment Facilities. He would leave Oak Ridge and worked at various DOE complexes across the nation until returning to Oak Ridge to help with the cleanup of the K-25 Site, which is now called the East Tennessee Technology Park (ETTP). Working for UCOR, an AECOM led partnership with Jacobs, his expertise has significantly helped the company work toward achieving Vision 2020, DOE's plan to complete ETTP cleanup by 2020. The site is being transformed into a private sector industrial park, national park, and conservation area. Conner is now semi-retired but serves as a consultant for UCOR.

588933 A Chemical Engineer in the Energy Journey

Dr. Quinta Nwanosike Warren, PE, PMP is a chemical engineer and an energy professional. She is an entrepreneur and has worked in industry, academia, and government. She holds a PhD in Chemical & BioMolecular Engineering from the Georgia Institute of Technology, and a Bachelors in Chemical Engineering from the Pennsylvania State University. She is a Project Management Professional, and a registered Professional Engineer.

Dr. Warren previously worked for ConocoPhillips for on CO2 capture research, and as a reservoir engineer for heavy oil and tight gas assets. She has worked with the US Department of Energy, learning about and contributing to domestic and international policy around carbon management and power generation. She also worked for the Millennium Challenge Corporation (MCC), a US government agency that works on international development. At MCC, she provided energy expertise and contributed to the development of energy projects.

Dr. Warren owns and runs a consulting company called Energy Research Consulting (www.EngrRC.com) which provides energy and water solutions and strategies. The company's projects include creating the STEM-focused Dakar American University of Science and Technology (www.DAUST.org) in Senegal, and designing a water treatment system for a mushroom farming initiative in Nigeria. Dr. Warren recently published a textbook titled "A Practical Guide to Oil & Gas Resource Characterization for Geologists and Reservoir Engineers."

588881 Using microcontact printing to mimic Cardiomyocyte regeneration

Cardiac diseases are the most prominent cause of death worldwide. Cardiomyocyte regeneration is imperative, due to the human body lacking in the ability to sufficiently regenerate cardiomyocytes (CMs). One-way cardiomyocytes can be regenerated is through differentiating human induced pluripotent stem cells(hiPSCs) into cardiomyocytes. In order to attach hiPSCs derived-cardiomyocytes in a manner that would mimic native conditions, various techniques can be used, such as microcontact printing. The microcontact printing technique allows for a specified geometry or pattern to be transferred onto the substrate that will be used for cardiomyocyte seeding. The pattern utilized in this study comes from the rectangular-pillared-array stamp. The preliminary results show that a general cell line of mouse embryonic fibroblasts (MEFs) can be constrained to this pattern.

588862 Direct contact membrane distillation (DCMD) membrane distillation techniques for desalination

Direct contact membrane distillation (DCMD) is one of the simplest and most widely used membrane distillation techniques for desalination. It is a thermally driven separation process and is sought to achieve high water vapor flux while being thermally efficient. Typically, a hydrophobic porous membrane is used and water vapor from the salt solution is allowed to pass through the pores and condense on the other side of the membrane. Investigation on other structures of membranes is carried out. Membranes include those in the form of an integral composite, in which the structures of PVDF membranes are changed, and a stacked composite, in which different membranes are stacked on top of each other (e.g. a hydrophobic ePTFE membrane over a hydrophilic PVDF membrane, and a hydrophobic PVDF membrane over another hydrophobic PVDF membrane). These membranes were found to yield a higher water vapor flux than the conventional single hydrophobic membrane. Membranes are also characterized using a variety of characterization techniques: Contact angle, liquid entry pressure (LEP), bubble-point pressure, scanning electron microscopy (SEM), and atomic force microscopy (AFM).

588882 Effects of a functionalized CNC on the rheological properties and printing behavior (Direct Ink Writing (DIW) of an epoxy resin

In this work we characterize the effects of a functionalized CNC on the rheological properties and printing behavior (Direct Ink Writing (DIW)) of an epoxy resin. CNC functionalization is achieved by grafting an activated medium chain fatty acid to substitute the hydroxyl functional groups. The level of CNC modification is evaluated using Fourier Transform Infrared (FTIR), thermo-gravimetric analysis (TGA), X-ray diffraction (XRD), and solvent dispersibility studies. Functionalized CNCs are dispersed into the epoxy resin (EPON 828) at loadings up to 20 wt%. Initially, rheological studies are used in this study to assess the printability of the inks. The rheological percolation threshold is the critical concentration at which there is a transition from liquid-like to solid-like behavior. For this system, this value was found to be at ~10wt%. We found that for the final printed structure to exhibit structural integrity, it is necessary to print at loadings slightly above this value. Future work will include microscopy (SEM and polarized light) of both printed and cast samples to observe the dispersion and alignment of the CNCs within the polymer matrix. Finally, the mechanical properties of all samples will be measured to show the effect of the reinforcement in the matrix.

588883 madison.garner1@my.hamptonu.edu

Conventional lithium batteries using an organic polymer electrolyte, exhibit several safety issues including leakage, poor chemical stability and flammability. To address these safety concerns the idea of a solid electrolyte has emerged. In the lab we specifically focused on the solid garnet-type electrolytes. The solid electrolyte of $Li_{6.28}Al_{0.24}La_3Zr_2O_{12}$ and $Li_{7.28}Al_{0.24}La_3Zr_2O_{12}$ (LALZO) powder were successfully synthesized by a co-precipitation method. LALZO pellets were then sintered at 1100 °C at different sintering times of 4,8, and 16 hours. With the lithium content of 7.28 per formula, impurity of La_2O_3 was obtained at the sintering time of 4 hours. When the sintering time increased to 8 and 16 hours, in addition to the La_2O_3, $La_2Zr_2O_7$ phase was obtained. When the lithium content decreased to 6.28 per formula, impurities of La_2O_3 and $La_2Zr_2O_7$ were obtained. The lattice parameters are all similar for different lithium content and different sintering times and are within the range of 12.93-12.97Å.

588874 Epitranscriptomics of RNA oxidation reveals mechanisms of lung toxicity and disease

Chemical modifications of proteins and nucleic acids play critical roles in the regulation of gene expression. New evidence suggests that environmental stresses may be involved in the misregulation of the functional activities of RNA modifications and pathways during lung distress and disease. Air pollution is the major source of environmental toxicity to the pulmonary cells. Components of air pollution can generate free radicals and perturb redox homeostasis in cells, leading to formation of the prominent RNA oxidative modification 8-oxo-7,8-dihydroguanine (8-oxoG). This mark is predominantly induced in respiratory cells, and it leads to genotoxicity as it preferentially mispairs with adenosine. Given the prevalence of RNA oxidation under environmental challenges, it is critical to identify novel molecular mechanisms involving RNA oxidation toxicity to cellular physiology and disease. To address this, we developed a high-throughput sequencing method to profile 8-oxoG in mRNAs of human bronchial epithelial BEAS-2B cells. Cells were exposed for 1.5 hours to realistic air pollution mixtures derived from the reaction of ozone (100 ppb) with acrolein (100 ppb), methacrolein (97ppb) and a-pinene (44ppb), common organic volatile compounds in the atmosphere. These compounds formed a multi-component gas-phase mixture and submicron secondary organic aerosols (PM 1.0) with concentrations ~ 40 $\mu g/m^3$. Using this approach, we found 42 transcripts that are consistently oxidized by air pollution. These transcripts belong to key oxidative stress, signaling, and metabolic pathways. Importantly, one of the mRNA transcripts that is more susceptible to oxidation is FDFT1, a protein that catalyzes a regulatory step in cholesterol synthesis and free-radical scavenging. Moreover, we confirmed a ~2-fold (p-value < 0.05) decrease in expression of FDFT1 protein, and a ~1.7-fold (p-value < 0.001) decrease in cellular cholesterol levels. To further validate the involvement of misregulation of cholesterol on cellular toxicity of air pollution, we knocked down FDFT1 using anti-sense siRNA in BEAS-2B cells. We found that FDFT1 knockdown promotes distinct morphological phenotypes typically observed in studies of lung stress and inflammation. Overall, our findings indicate that air pollution influences the formation of 8-oxoG marks in transcripts of epithelial lung cells leading to alterations on cellular function and cell morphology.

588871 Cell-to-cell variations define age-dependent cell migration through fractional redistributions among motility states

As humans age, cells within organs and tissues undergo profound biophysical and biomolecular changes, which significantly influence the rate of progressive functional decline. Recently we have demonstrated that aging information is encoded within biophysical properties of cells, and as such can be used as biomarkers of aging in health and disease. The coordinated movement of cells within organs and tissues are important in the contexts of both homeostasis and disease; such as in the surveillance of immune cells and during cancer metastasis. Although it is now recognized that there are ensemble decreases in the motile capacity of cells with increasing age, it is unclear how this motile regulation becomes impaired. To address this question we procured a panel of primary dermal fibroblasts from donors aged 2-92 years old, and tracked the movement of cells seeded on collagen-coated substrates for ten hours, taking images every 3 minutes. Quantification of bulk motility properties using the anisotropic persistent random walk model, revealed that there was indeed a decrease in the overall capacity of cells to move with increasing age (including displacement, directionality and velocity). Looking further into these findings, we plotted the trajectories of individual cells across all ages, and found that there was a high decree os cellular heterogeneity among cells, with some cells from elderly donors exhibiting highly-motile features, and vice versa. Prompted by the extent of cell-to-cell variations, we hypothesized that these age-dependent decreases in cell motility was not due to global decreases in the movement of all cells, but a re-distribution of the proportions of highly-motile and non-motile cells with increasing age. Taking advantage of the single-cell nature of the data, we compiled the trajectories of cells across all ages and computed eight motility parameters that described the spatiotemporal movement; e.g. velocity, displacement, anisotropy and diffusivity. Then from this data matrix we used unsupervised hierarchical clustering and identified eight clusters which defined 'motility states' exhibiting distinct spatial patterns, as indicated by the trajectories of individual cells within each cluster, and t-stochastic neighbor embedding analysis. Next, we asked whether certain clusters had increased propensities for cells of a particular age. Quantifying the abundance of cells classified per cluster for all ages, we found that there was an increased propensity of various cluster for young and elderly donors, however, for post-adolescent/adult donors there was a similar likelihood of either cluster. This finding was interesting as it points to increased heterogeneity during the 'middle-age' period. To further verify this result, we quantified the 'activity' of cells per age, as determined by the 1- dimensional displacements per cell based on the frequency and magnitude of displacement bursts. Utilizing a similar clustering approach as before, we identified five clusters that defined the activity of each cell across all ages. interestingly, while the temporal patterns of the bursts were different among groups, plotting the trajectories of individual cells we were unable to identify distinct spatial motility patterns. However, plotting the abundance of each 'activity' cluster per age, showed a less-defined but similar tendency in cellular heterogeneity. Combining the information regarding the spatial-temporal and activity per cell, we computed the abundance of cells within the 40 motility states (8 spatio-temporal clusters x 5 activity clusters) to define motility landscapes, and asked how this differed as a function of age group. Interestingly, we found that there was a clear difference among age groups regarding the propensity for certain states, with the significant gain and loss of states across ages. Quantifying the Shannon entropy of each landscape, we confirmed the results of a maximization of heterogeneity for donors classified as post-adolescent/adults.

588870 Microfluidics for the Engineering Transport Classroom: Multipath Flow

Microfluidic devices can be utilized to recapitulate a myriad of engineering principles. While polydimethylsiloxane (PDMS) chips are the most common, hybrid polyethylene terephthalate laminate (PETL) microfluidic devices chips have shown promise in being cheaper, more affordable, and faster alternatives to the current "gold-standard". In an effort to apply newer, more hands-on methods to assist in teaching engineering principles in the classroom, PETL devices were successfully implemented in modeling multipath flow in microfluidic systems. Future developments may allow for more complex means of integrating other engineering concepts for higher accuracy and a wider scope of application.

578803 Solvent Absorbent Amphiphilic Microgels for Pickering Emulsions

Pickering emulsions are critical to systems where encapsulation of an oil or water phase in an immiscible medium is required. These coated emulsions can harbor oils in water or even water in oils. While hard particles are traditionally used, microgel particles have also been used as emulsion stabilizers. These polymer colloids differ from rigid particles in many ways. When compared to microgels, colloidal particles are lyophobic, meaning they need surface modification to be preferably "wet" by a solvent. Microgels can be made to be lyophilic, having an attraction to the solvent that swells them. Microgels are cross-linked polymer networks that absorb and desorb solvents where the type of solvent used to swell a microgel can be tailored during particle synthesis.

Using a Polydimethylsiloxane microfluidic device, via soft lithography, the flow streams of two separate polymer solutions were brought into contact to form Janus droplets. The individual droplets were then exposed to UV light to induce cross-linking via photo-polymerization to form monodispersed microgel particles with sectioned hydrophilic and hydrophobic sides. 45 wt% Polyethylene Glycol diacrylate (PEGDA), 5 wt% ethoxilated trimethylopropane triacrylate (ETPTA), 1 wt% 2-hydroxy-2-methyl-propiophenone (HMP), and 49 wt% water were mixed together to form the hydrophilic portion while 99 wt% Polypropylene Glycol diacrylate (PPGDA)) and 1 wt% 2-hydroxy-2-methyl-propiophenone are added together to form the hydrophobic side. In crosslinking these two systems together, we form single particles that have one side capable of absorbing water and another with the ability to absorb and desorb organic solvents. These particles were used to make stable water in oil and oil in water Pickering emulsions.

PEG functionalized CdSe/ZnS core-shell type quantum dots in water and CdSeS/ZnS alloyed quantum dots in toluene were used to observe the absorption abilities of the hydrophilic and hydrophobic microgel sides respectively with a confocal microscope. While oriented at the water/oil interfaces, the respective microgel sides swell and capture the dispersed nanocrystals. These amphiphilic microgels pose as an interesting tool in stabilizing emulsion droplets for cosmetic, pharmacy, and petroleum industries.

588868 Determining the Effects of Adipose-derived Exosomes on Stem Cell Proliferation and Adipogenesis

Determining the Effects of Adipose-derived Exosomes on Stem Cell Proliferation and Adipogenesis

The field of tissue regeneration has made many strides to improve therapeutic methods and restore missing functions to patients with the introduction of stem cell therapies. Transplantation of fat tissue has provided beneficial metabolic effects to patients by supplying sufficient vascularization and innervation[1]. Despite the benefits that adipose tissue therapeutics provide, there persists limitations such as the restriction to autologous therapies due to immunogenic responses. Exosomes may serve as a novel cell-free method to provide the same therapeutic properties while circumventing potential immunogenic responses. This is beneficial to patients who are deficient in fat content and would not have tissue for an autologous therapy, such as lean and pediatric patients. The purpose of this project is to determine the therapeutic potential of adipose-derived exosomes on tissue regeneration and their influence on human adipose-derived stem cell (hASC) adipogenesis.

Fig 1. Schematic of growth factor effects on adipogenic gene expression[4]

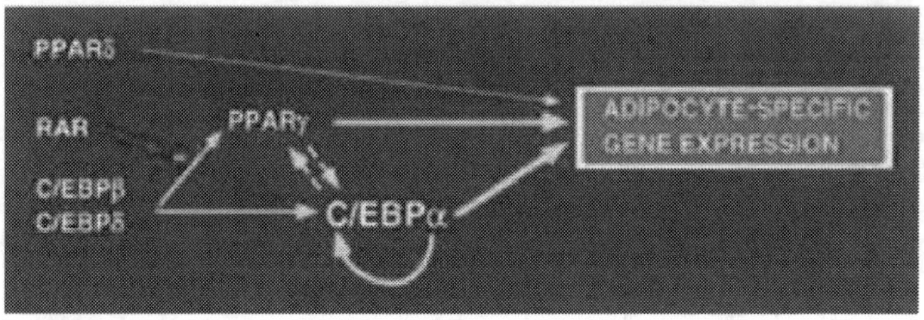

Exosomes are small vesicles (40-100nm) that are secreted by a variety of cell types into the extracellular matrix. Within an apoptotic cell, the intracellular matrix fuses with the plasma membrane to produce multivesicular bodies (MVB). Through fusion with the plasma membrane, exosomes retain the properties that are associated with the native cell. These vesicles have surface ligands that help to facilitate cell-to-cell interactions without the need for direct cell contact. By inducing the expression of target growth factors, we can interact with the various surface proteins that are involved in cell proliferation and differentiation[2]. Exosomes fuse with target cells to exchange membrane proteins and cytosol to signal intended functions[3]. The surface proteins of cells provide a more accessible route for communication than internal targets. Transcription factors that have been found to induce adipogenesis include members of the CAAT/enhancer binding protein (C/EBP) and peroxisome proliferatoractivated receptor (PPAR) families (Figure 1). Through a combination of interactions, each transcription factors can directly or indirectly activate the expression of adipogenesis. C/EBPß and C/EBPd have synergistic effects to influence PPARγ and C/EBPa, each of which have a direct influence on gene expression. PPARd also has a direct influence on adipocyte gene expression, with slightly less significance[4].

By directing the surface interactions, it is possible to have a cell perform specific tasks, in response to the given environment. **I hypothesize that adipose-derived exosomes can serve as a therapeutic agent by promoting proliferation and differentiation.**

Currently, exosomes from differentiating human skeletal muscle cells have been tested to observe the regenerative properties within skeletal muscle tissue. Studies in other tissues have also shown that mesenchymal stem cells (MSC) derived exosomes have functions such as repairing tissue damage, suppressing inflammatory damage, and modulating the immune system[5]. Working with adipose-derived tissue, I want to determine the therapeutic potential that exosomes possess. Future work will consist of determining the mechanism behind each function. This therapy has the potential to assist with tissue regeneration in patients that are dealing with adipose tissue defects. Exosomes show promise for transferring regenerative therapeutics between patients without having to encounter immunogenic rejection, a major setback when dealing with stem cells. This research will open new avenues for dealing with patients who lack the tissue mass for autologous regeneration therapeutics such as lean or pediatric patients. Restoring functions to damaged organs can one day be achieved by mediating the repair of the desired tissue through exosome mediated cell-to-cell signaling.

588867 Predicting Cancer Progression using a Population Balance Model and Supporting Evidence from Zebrafish Melanoma Studies

Cancers are complicated diseases that in 2018, killed 600,000 people in the US. Reliable models for the progression of the disease save lives by guiding treatment and identifying risk factors. Of special interest is the perplexing phenomena of tumor dormancy and recurrence, where after treatment or surgery, patients remain disease-free for years or decades only to experience a sudden relapse of the same disease. We seek to explain this and other phenomena with a simple population balance model that describes how growth effects (e.g. mitosis), reduction effects (e.g. immunity or treatment) and metastases (inception of new tumors) determine disease progression.

In the model, these effects are represented as tumor-size-dependent rate parameters in a stochastic process. Treating cancer growth as an advection-diffusion transport process in tumor size space allows us to solve for the size distribution of tumors in a population of patients. Simulations using the model show that tumor size can change diffusively for long times after which the tumor begins to grow or shrink, indicating the model recapitulates the phenomena of dormancy and recurrence without any manipulation.

We used the zebrafish melanoma model to quickly acquire tumor growth data under various conditions including growth with and without the natural immunity and growth in different genders. The size-dependent parameter approach described experimental data well. We also saw statistically different immunity parameters between male and female fish, indicating how the model can be used to identify and elucidate population differences in disease progression. We further show through simulations backed by experimental data how changes in parameters affect the possibility of relapse using both parameters from animal and human cancers.

The model demonstrates a mechanism for tumor dormancy and recurrence that does not rely on any specific biological explanation such as the so called 'angiogenic switch.' Rather it shows that a series of small, incremental changes such as those due to a size-dependence can generate dormancy and recurrence. Recognizing that these phenomena could have no single biological cause mandates a radical shift in research methodology for this problem.

588866 Research Explore Removing Zinc Oxide from the coastal marine environments prone to excessive sunscreen runoff.

With recent highlights of the state of coral reefs and help from social media, the public has finally become aware of the adverse consequences of the widely used chemical, Oxybenzone, in conventional sunscreen, as well as its direct effect on coral bleaching. The issue has become so prevalent in coastal areas, that a prohibition of Oxybenzone will take action in Florida and Hawaii in 2021 (Calderone, par. 1-2). To support the movement, consumers have been taking it on their own behalf to choose "coral reef safe" sunscreens. The two prominent ingredients in the eco-friendly sunscreen movement are Zinc Oxide and Titanium Dioxide. Recent studies, however, point to evidence that these two chemicals may not be as Eco-friendly as producers are suggesting.

Zinc Oxide in sunscreen has a tendency to dissolve into Zinc ions in water, and with decreasing the particle size, there is an increase in the likelihood of releasing free ions. The effects of Zinc ions in water has been shown to promote the growth of algae. Additionally, ZnO showed increased effects in solutions that were aged, suggesting that the true effects of the chemical won't be apparent until later on.

On the other hand, initially, it may seem intuitive to pursue Titanium Dioxide sunscreen. Unfortunately, Titanium Dioxide is a 2B carcinogen, meaning that while it is not a confirmed fact yet, it may cause cancer to humans (Smijs, par 21).

Additionally, new generation organic sunscreens, while they have been on the market for the past few years in the EU, cannot be used yet as the compounds they contain have not yet been approved in the U.S. by the FDA.

588865 Spray Drying of Griseofulvin Nanosuspensions and Solutions for Preparation of Nanocomposites and Amorphous Solid Dispersions: Comparative Assessment of Drug Release

BCS class II drugs are compounds that are poorly water-soluble and exhibit high permeability. In order to increase the dissolution rate and bioavailability of these compounds, two specific processes can be used: amorphous solid dispersions (ASD) and nanocomposites. ASD's disrupt the drug crystal lattice which causes an increase in the saturation solubility, whereas nanocomposites exhibit very large surface areas; both of which cause a faster dissolution. The goal of this project is to compare the release of the active pharmaceutical ingredient Griseofulvin (GF) from nanocomposites and that from ASD's, both having identical formulations.

GF nanocomposites were prepared by wet media milling (nanomilling) an aqueous GF suspension followed by spray drying. Two different polymers, hydroxypropyl cellulose (HPC) and Soluplus, with an anionic surfactant, sodium dodecyl sulfate (SDS), were used for the stabilization of drug nanoparticles. ASD's were formulated by spray drying the respective drug-polymer in acetone-DI water. The suspensions were characterized via laser diffraction, while the spray-dried powders were characterized via laser diffraction, X-ray diffraction (XRD), and dissolution testing. XRD diffractograms confirmed that for both polymers, spray drying of the aqueous nanosuspension led to nanocomposites, whereas that of the drug solution led to ASD. At supersaturating conditions, drug dissolution from ASD was faster than that from the nanocomposites.

588864 Properties of Modified Kraft Lignin for Use in Thermoplastic Composites

With the ever-increasing use and popularity of thermoplastics, effective waste management continues to be an area of interest. In order to preserve or improve the quality of the recycled material, thermoplastic composites have been researched. The properties of modified Kraft lignin by two methods (i.e. acetylation and maleation) are investigated and the results are compared with virgin Kraft lignin. Kraft lignin is considered a commercial byproduct of the Kraft pulping of wood chips. In this study, the thermal and morphological properties of the esterified Kraft lignin are analyzed. The lignin is modified by esterification using acetic and maleic anhydride to alter solubility behavior in nonpolar solvents for application in thermoplastics. The morphology of the composite materials is examined using scanning electron microscopy (SEM). Differential scanning calorimetry (DSC) and thermogravimetric analysis (TGA) is used to study the glass transition temperature and thermal stability of the modified lignin samples. The results are compared with virgin Kraft lignin.

588877 *2019 Eminent Engineer Recipient:* Ana Davis, Syngenta: "What I know now that I didn't know then – A Chemical Engineer's Journey"

Ana will reflect on her journey as chemical engineer and touch on the "breakthrough" moments that have led to positive changes in her life and career as a chemical engineer. She will touch on topics such as Diversity, Leadership, Empowerment and Courage. She will share personal stories that have helped her grow as a leader and as an individual.

588878 *2019 Eminent Engineer Recipient:* Zenaida Otero Gephardt, Rowan University - "Talent and Hard work are essential for success"

Talent and hard work are essential for success, but without mentors and meaningful support, success can elude even the most talented and industrious among us. This is especially true for those who are underrepresented and have historically lacked access to opportunity. The presentation will focus on the importance of mentors throughout the presenter's career and on how mentoring is an essential investment in the chemical engineering profession and in the global community. As mentors, we see reflected the best of ourselves and continuously improve in our professional and personal lives regardless of career stage. The presenter will also discuss her personal experience with the type of support that is necessary to overcome poverty and disadvantage. Mentoring is key to the broadening of participation in the profession that isessential to identify opportunities in future challenges, and to implement solutions that improve quality of life worldwide.

588880 *2019 William W. Grimes Award:* Dr. Christine Grant, NCSU- "Mentors, Mentors Everywhere: Weaving Informal and Formal Mentoring into a Robust Chemical Sciences Mentoring Quilt"

In an era of convergence, a clear understanding of the discipline-specific cultures is at the root of interdisciplinary mentoring. Informal and formal mentoring manifests differently for women and underrepresented minorities in the chemical sciences where more work is required to achieve parity, equity, and inclusion for diverse groups. This is particularly true for Women of Color (WOC) who occupy a unique place at the intersection of race and gender in STEM fields. Intentional, informed, mentoring includes informal interactions at formal conference events, where cross-cultural mentoring can result in robust, long-lasting mentoring teams.

580292 Background and Discussion about Retaining Women Engineers

Dr. Urmila Diwekar is currently President of the Stochastic Research Technologies LLC, and the Vishwamitra Research Institute (VRI, www.vri-custom.org), a non-profit research organization that she founded in 2004 to pursue multidisciplinary research in the areas of Optimization under Uncertainty and Computer-aided Design applied to Energy, Environment, and Sustainability.

From 2002-2004, she was a Professor in the Departments of Chemical Engineering, Bio-Engineering, and Industrial Engineering, and in the Institute for Environmental Science and Policy, at the University of Illinois at Chicago (UIC). From 1991-2002 she was on the faculty of the Carnegie Mellon University, with early promotions to both the Associate and the Full Professor level.

In chemical engineering, she has worked extensively in the areas of simulation, design, optimization, control, stochastic modeling, and synthesis of chemical processes. She has made major contributions to research on batch distillation, and this work is well recognized. She is the author of more than 165 peer-reviewed research papers, 7 books, and 15 chapters, and has given over 350 presentations and seminars, and has chaired numerous sessions in national and international meetings. In separations, she is well-known for her pioneering work in batch distillation design, simulation, optimization, and optimal control.

She has published more than 31 peer-reviewed journal papers, 65 presentations (15 plenary/keynote presentations), and conducted 11 educational workshops on this topic. She has also published the first book on this topic. This is an authoritative, first of its kind book which is widely used all over the world. This book received excellent reviews and the second edition of this book was published in early 2012. She is the author of a comprehensive software package for batch distillation simulation, optimal design, and control called MultiBatchDS which is used in pharma and specialty chemical companies in USA, Europe, and Asia.

In this session, she will give her personal opinion about how to maintain women in the industry.

580293 Background and Discussion about Retaining Chemical Engineering Women in Industry

Gayle J. Gibson led the DuPont Engineering Research and Technology organization of internal engineering consultants and the company's Field Engineering Program for early-career engineers.

Gayle joined DuPont in 1983 and held a variety of roles through which she implemented best practices and enabled global business revitalization on behalf of multiple, high-performing business units. Her contributions included engineering design, process engineering, research and development, product and process development, manufacturing supervision, market development, business strategy, growth and supply chain improvement.

From 2008-2013, Gayle was Director – Corporate Operations for the Chair of the Board of Directors and Chief Executive Officer of DuPont. In that role she facilitated processes and engagement among key DuPont operating and leadership teams. This included working across businesses, functions and regions to drive alignment and clarity on priorities, advancing core values and resolving issues in line with the company's mission and strategic objectives. She also supported the Chief Executive Officer as co-lead of the National Academy of Engineering project "Changing the Conversation – from Research to Action" to improve the public understanding of engineering.

A certified Six Sigma Champion, Gayle holds a Bachelor of Science in chemical engineering from Texas A&M University and a Master of Business Administration from the European University. She is a member of AIChE, Tau Beta Pi and the Society of Women Engineers

In this session, she will be giving her personal thoughts on how to retain women engineers in the industry.

580291 Background and Discussion about Retaining Women in Industry

Ana Davis heads Health, Safety, and Environment (HSE) in North America for Syngenta, a global agriculture company. She manages all of Syngenta's corporate HSE programs, which include active ingredient manufacturing, product formulation, seeds operations, and commercial sales. Her team develops tools to improve safety culture across the company.

Ana began her career as a process control engineer with Exxon, later working as a facilities engineer for Advanced Micro Devices. Prior to joining Syngenta in 2009, she held a variety of production and HSE management roles with Kao Specialties Americas. Before that, she was a plant manager at Dow and held several positions with Union Carbide Corp., becoming a plant production manager after the Dow Chemical merger.

A native of Spain, Ana was recognized in 2005 with the Hispanic Engineer National Achievement Award for Professional Achievement in Industry. She serves on the advisory boards of AIChE's Center for Chemical Process Safety (CCPS) and North Carolina State Univ.'s Dept. of Chemical and Biomolecular Engineering. Ana received her BS degree in chemical engineering from North Carolina State Univ., where she and her husband have established an endowment for chemical engineering undergraduate students.

In this session Ana will give her personal point of view about maintaining women engineers in the industry.

587477 Background and Discussion about Retaining Women Engineers in Industry

Bill serves as Director – Engineering Technology for the Chemours Company. He is accountable for leading the company's efforts to transform and renew the leveraged engineering competencies in support of Manufacturing Operations, large Capital Growth Projects, Sustainability and Process Safety.

Bill has over 30 years of experience with DuPont and Chemours in R&D, Engineering and Manufacturing technology management, as a well having assignments in corporate risk management and trade secret protection.

Bill graduated with a B.S. degree in Chemical Engineering from Columbia University and received his M.S.C.E.P. and Ph.D. degrees in Chemical Engineering from the Massachusetts Institute of Technology.

AUTHOR INDEX